Finding Myself in the Storm

BRENDA F. KEITH

Copyright © 2023 Brenda F. Keith
All rights reserved
First Edition

Fulton Books
Meadville, PA

Published by Fulton Books 2023

ISBN 979-8-88982-588-3 (paperback)
ISBN 979-8-88982-589-0 (digital)

Printed in the United States of America

This book is dedicated to my children, Sharette Rachon Keith-Foster, Deitrich Dewillis Keith, and Larry Donnell Hamilton Jr., and to Sharette for encouraging me to write this book and for loving and supporting me along the way. May God keep blessing you physically, spiritually, mentally, and emotionally.

To my grandchildren, Javoughn Lance Perkins, Deitrich Keith Jr., KeNia Webber, Emile Foster III, Jordan Foster, DeShaun Keith, Dylan Keith, and Jeffery Foster, for being the loves of my life, and may God bless you.

To my sisters, Betty Ann Chancellor, Marie Malone, and Yvonnia Chancellor, for being there for me throughout my journey. May God keep you in His hands.

Emile the second, this is my son in law. Emile Foster II and Nyra Webber for your kindness and generosity and for being a part of my family.

In loving memory of Wylinsky Keith, my beloved husband; Fred (Jack) Chancellor, my father; Lue Edna Cargle-

Chancellor, my mother; Annie Mae Thomas and Georgia Ann Baker, my sisters; and Fred Chancellor Jr., Tommy Gene Chancellor, and Denson Chancellor; my brothers.

ACKNOWLEDGMENTS

I thank God for blessing me physically, spiritually, mentally, emotionally, educationally, financially, and socially. What I mean by physically is He has kept my body safe from abuse and physical harm; spiritually, He taught me how to love Him first, myself, and then others and to care about people, animals, and the earth. Mentally, He taught me how to be humble and protect my mind. Emotionally, He taught me how to stay in control and be responsible for the choices I make. Educationally, He allowed me to go back to school and get a degree.

Financially, with that degree I landed a well-paying job to bring me out of a lower-level status to a middle-class status; and socially, He taught me how to communicate with others intellectually.

I thank my mom for doing all she could to raise me the best that she knew how considering she had health problems starting from as early as I can remember and the limited resources the family had.

I thank my sister Betty for helping my mom take care of me, for buying clothes and other necessities for me when my mom and dad couldn't afford it, and for opening her door to me in a time of need.

I thank my late brother Fred Chancellor Jr. for bringing me to town on Fridays after work buying whatever snacks I wanted when I was ten years old and for buying me a shiny new bike for Christmas when I was twelve.

God blessed me with two sisters: Marie Malone and Yvonnia Chancellor, and I'm grateful to have had them by my side throughout my life. I'm also grateful to have had four other siblings: Annie Mae Thomas, Tommy Gene (Clay) Chancellor, Georgia Ann Baker, and Denson Chancellor who have all left us and are resting in peace.

I want to give thanks to three of my teachers who played an important part of my journey: Mrs. Brewer, Mrs. Overton, and Mrs. Limbaugh for believing in me and giving good advice.

Between the ages of thirteen and fifteen, I lost three family members: my oldest sister, then my oldest brother, and my mother. Then I met Larry Hamilton Sr., whom I though was a friend, but meeting him was like a fork in the road. I had to decide which path to take, so I took the one with him and found that this wasn't the right path.

I moved back to my hometown and met my second husband, Wylinsky Keith. This seemed like the right path, but it was cut short on that faithful day on September 1979 when there was a tragic accident. I moved from my hometown to New Orleans, Louisiana.

While living in New Orleans, Louisiana, I had a serious talk with my God, and He told me that to be happy, I needed to take a new road, and He revealed to me that this one should be the uncharted—a road that no one else in my immediate family has ever taken before—so I took that challenge.

I bought a new home, joined a church, and met a new friend, Lawrence Spain.

We were together for a few years, and then I went back to school. It seemed that the new relationship couldn't stand the signs of time, so we went our separate ways.

I graduated from the University of New Orleans the summer of 1995 with a bachelor's degree in early childhood education. I've found new happiness with my loving God, myself, and my family.

CONTENTS

Introduction xi

Chapter 1: From Toddler Years to Work and

                Play on the Farm 1

Chapter 2: Starting School and a Talk with My God 12

Chapter 3: A Move from Farm to Town 17

Chapter 4: Attending a New School and

                Meeting Friends 25

Chapter 5: Transfer to Another School 30

Chapter 6: Be Careful Who You Befriend and

                Missing Mom 36

Chapter 7: A Temporary Move to New

                Orleans, Louisiana 49

Chapter 8: My Knight in Shining Armor 54

Chapter 9: A Permanent Move to New Orleans
            Louisiana 61
Chapter 10: My Career Decision...................................66
Chapter 11: Looking Forward to Retirement.................71
Chapter 12: Living the Life My God Designed for Me ...75

Epilogue..81

INTRODUCTION

From a sharecropper's daughter to an elementary school teacher, I was born in the very small town of Vincent, Alabama. It's about 36.2 miles east of Birmingham.

I don't know what the population was then, but in 2019, it was only 1,918. There weren't any other houses in close to us except the one my oldest brother lived in. We had to walk or drive at least two miles before we could see another house.

There weren't any traffic lights, only signs and one flashing yellow light at the main intersection of what we called downtown. We did not have any major grocery stores, only small markets and gas stations. To shop at a major grocery store, my family had to drive to other towns that had a bigger population than Vincent to shop at the major stores.

From conception to around one and a half years old, I had zero knowledge of self. When I was around two years old, I remember my brother Tommy Gene (Clay) Chancellor used to pick me up and lower me down very fast as if I was on a roller coaster. Each time I went up, my head would bump the ceiling because he was very tall. This is how he played with me. I remember the last time he played with me that way was when one day he lifted me up, but when I came down, I landed in a tub of hot ashes. My mom said he had been drinking and didn't realize the ashes were hot. Someone had cleaned the ashes from the fireplace and left the tub on the floor in the living room. He wasn't allowed to play with me that way again.

We lived on a farm in a three-room cabin with Mom, Dad, and seven children living there. No, it wasn't three bedrooms, it was three rooms: a living room, a bedroom, and a kitchen. So there was a total of nine people living there; it would have been eleven, but the eldest sister, Annie Mae, who was about eighteen years old at the time married and moved to Wilsonville, Alabama, with her husband, Henry Thomas. They were only about twelve miles up the road from the family. At sixteen, the eldest brother moved with his girlfriend to the next cabin that was about fifty feet away from the family. I was the youngest of the children. Back then, eighteen and sixteen were considered as adults. Even though some may have considered them as adults, from my point of view, they were still children.

There wasn't any running water inside the house. We had a well outside and an outhouse. The well was very unsafe because there wasn't any structure; there was once a structure, but it no longer existed. Someone covered the hole with two pieces of plywood, but we still would draw water with a bucket on a rope.

I can remember my dad and brothers would ploy the field with a tractor next to the cabin for Mom and all the sisters to sow vegetables seeds, and Mom would plant sunflowers, roses, and other beautiful flowers in the front yard.

This was fun for me; I may have been about three or four years old at the time.

I had lots of fun on the farm from planting seeds, picking vegetables, feeding the chickens, gathering eggs, and feeding the pigs. I also enjoyed playing with my sisters and youngest brother. We played hunting with sticks and brought along three dogs for help.

At this time in my life, I had a bit more knowledge and started really observing my surroundings. I noticed my dad's behavior was changing. He would drink alcohol and became physically abusive to Mom. The family lived with that.

I started first grade at Vincent Elementary School at age seven. Because my birthday is November, I had to wait another year. School was great until I entered fourth grade; some of the classmates weren't so nice. We lived on the farm until I was twelve, seeing more instability within the family. Rules were not established as they should have been.

During my twelfth year on the farm, we moved from Vincent, Alabama, to another small town about seven miles east of Vincent named Childersburg, Alabama. Childersburg's population was approximately one thousand more than Vincent, Alabama; it had about two signal lights. We were happy there for a short time. My mom, dad, youngest brother, and I joined the neighborhood church called Enon Baptist Church. I started school and met new friends, and I lost two family members. Also I married at seventeen and did some travel.

Unfortunately, the marriage didn't work, so I filed for a divorce. About a year later, I met my second husband and bought a home in Childersburg. We were a happy family until the unfortunate happened again, but this time, it wasn't a divorce; my husband had a tragic accident in September of 1979.

In 1981, my children and I moved and settled in New Orleans. I went back to school and ended up getting an associate degree in business and a bachelor's degree in education. After twenty-five years of teaching, I'm now retired and enjoying life with my children and grandchildren.

Although I grew up in a dysfunctional family and was in a dark place, I don't want anyone to feel sorrow for me now. Back then is when I needed support. I've lived the past, and God gave me the structure and resources I needed to rise above the situations. Life is like a *fork in the road*; it comes to a point or situation when one is forced to decide which road to take. God will allow you to travel all roads because He knows you'll come back. During the early parts of this journey, I was undecided about which road to take, the left or the right. So I turned my life over to God, and He didn't allow me to take the road left or right. He allowed me to chart a new road for myself and others who wanted to follow. The road I took was the uncharted, and this is my story.

CHAPTER 1

From Toddler Years to Work and Play on the Farm

I was born on November 06, 1954 to Fred (Jack) Chancellor from Shelby County and Lue Edna Cargle from Talladega County both in Alabama. Fred (Jack) Chancellor was my father's name but was called *Jack* since the oldest brother's name was Fred Chancellor Jr.; not to get their names confused, but *Jack* was the name everyone called him, and they called my mother Edna.

My father was from a family of fifteen siblings, and my mother's family was from a family of ten siblings. The two joined together at very early ages and had a total of nine siblings: six girls and three boys. The eldest sister was Annie Mae, then Betty, Marie, Yvonnia, and me. The eldest brother was Fred Jr., then Tommy Gene (Clay), and Denson. I'm the youngest of them all.

I was born in a three-room cabin on a farm. I shared these three rooms with my mother, father, and seven siblings. I say seven because the eldest sister married at eighteen and moved away with her husband before I was born. My mom gave birth to me at home as she did with all her children. My dad was a sharecropper on Mr. Lee Chancellor and his sister Patsy Chancellor's farm, and his brother A. G. (Tot) Chancellor lived on the farm also. He lived in a two-room cabin, which was about one-half mile from the family's cabin, and he worked on the farm also. My dad's surname is the same as Mr. Lee and Ms. Patsy Chancellor because my great-grandparents were slaves of their ancestors during slavery, and that's the name my family was given.

My mother and father didn't get very far in school; they had only an elementary education, but they learned a lot from their experiences and environment. During my early childhood years, Dad worked on the farm, and Mom cleaned homes. The two eldest brothers worked in pulpwood. We didn't have as much as some others as I was growing up, but we seemed happy.

I really loved my mom and wanted to go with her everywhere she went. I liked being carried on her hip. I remember when I was three years old, my brother Fred Jr. had moved into a two-room cabin with his girlfriend a few steps up the path. Mom and I were going to visit them, and she carried me on her hip. I guess I was getting heavy because I remember very vividly it had rained, and we came upon a puddle of water. It was across the path, and we couldn't go around it, so Mom put me down on the ground. She had long legs, so she stepped across the puddle. Then she told me to across, but I was scared and didn't move. It looked like a river to me. I was only three years old. Then she told me if I didn't walk across the puddle, the *bogeyman* would get me, so I ran through the water. From that day on, she hardly carried me anywhere again.

There were lots of things to do on the farm like planting seeds and harvesting vegetables, to feeding the pigs, chicken, and gathering eggs. My favorite job was gathering eggs. Dad had built a chicken coop in the backyard with long shelves and straw for the chicken to lay their eggs. It was shaped like a small house and was big enough to walk inside. It even had a door to keep the foxes and other animals away from the chicken. Foxes, wolves, and coyotes were prevalent around there. But the dogs helped to keep them away. Gathering eggs were like hunting Easter eggs. Sometimes I would find some with a soft, clear shell; you could see the yoke inside without cracking the egg open.

I used to go to the barn with my mother, which was up the road on Lee Chancellor's property. She used to go there to milk the cows. The cows, horses, and goats belonged to Mr. Lee. The family helped maintained the farm for a profit of its goods, but we had a few pigs and a whole lot of chicken. One day, Mom made a mistake milking a cow; she sat behind the cow instead of beside her, and the cow kicked her; fortunately, she was okay.

Sometimes I would see my dad feeding the horses and cows some hay with the pitchfork. The horses would get an extra treat of dry corn on the cob. I would see him cutting grass in the fields with the tractor; sometimes he would let me sit in his lap and stir the wheel. He would let the older ones cut the grass all by themselves without him being on it. Those were the good old times.

Mr. Lee had apple and pear orchards down by the river. He allowed the family to get whatever fruits we wanted. I would go with them sometimes and not only did we get apples and pears, but we would come back with wild fruits, such as blueberries, blackberries, and plumbs. The wild plumbs were a little different than the ones bought in stores; they were sweeter and had more juice. We had a couple of fruit trees in our yard at the cabin, a peach tree and a persimmon tree. There was another fruit we ate that grew on vines like grapes. They were called *muscadines* (Wikipedia, https://en.m. wikipedia.org>wiki). The covering was not as soft as the covering on grapes; instead, it's kind of tough or dense, but they tasted good.

When we needed vegetables, we would harvest some from the garden. There was a wide variety to choose from, such as corn, okra, cabbage, all others in the green's family, beets, radish, tomatoes, peppers, sweet potatoes, and white potatoes, just to name a few.

When we needed meats, it depended on what my parents preferred. If they wanted chicken, Mom and my sister Yvonnia would go outside and catch a couple since they were the two faster runners in the family. Then Mom would pop the neck, pluck it, and prepare it for cooking. If they wanted pork, my dad would slaughter a pig. That was a scene I just couldn't watch. Whenever that happened, I would go inside and cover my ears trying to block out the sound. My parents would clean and prepare it for use and then put it in the smokehouse and would have porkchops, ham, ribs, roast, ham hocks, and any other parts of a pig anyone would normally use, even the chitterlings.

Just to mention, there were other fruit trees on the land, such as honey locust, pomegranate, and pecan. Honey locust was on our property, and Mom used to make some type of brew with the young sweet pods. (Honey locus - Wikipedia, (https://en.m.wikipedia.org>wiki)). Another interesting thing we used to eat were wild onions, which are also called ramps. They looked like scallions but were smaller and more delicate; they have one or two flat, broad leaves.

A good thing I can say about the family is, although it was large, we never went hungry while on the farm. That was because of the crops and livestock. Besides that, we did visit the grocery store about once weekly to get some other food items and other product we needed. Also my dad and uncle world go hunting in the wood about once a month for wild games, such as fowls, rabbits, squirrels, deer, raccoons, and opossums. I did eat some of all those animals, except raccoons and opossums. My mother told me, when I was about six years old, to look under their paws; their feet look like baby feet. I checked it out, and they did look like baby feet. I don't remember eating raccoons, but I did eat opossums for about two or three times. First, the meat smelled funky when preparing it, and it tasted just as funky while eating it. Then I asked my mother why they smelled and tasted like that, and she said because they are scavengers. They devoir other dead animals, preferably those that has been decayed for some days. "Yuck!" That was it for me.

One day my sister Yvonnia and brother Denson brought me along with them to hunt rabbits with sticks and three dogs. Shane was a shaggy dog with reddish-brown fur. Snowball was shaggy and white with a little black on the tip of his tail and a black spot on his forehead. Trackdown was a little taller than the other two dogs with very short brown fur; he resembled a greyhound. We were trying to copy our

Dad and uncle, so we walked a very short distance from a field to the woods just about a half mile further down. Whenever we spotted a rabbit, it would run into the briar patch or bushes. The dogs would go in and run them out. Standing in a pyramid form around the bush, we waited until he came out and whopped him with the sticks. We would bring home as many rabbits as my dad Fred (Jack) and Uncle Tot did, but not that day.

I don't know what breed the dogs were; they were probably mixed breed, but my brother Fred Jr. used to put gunpowder in their food. He said, "It will make them mean." We were looking for rabbits that day, but the dogs spotted a goat in the woods and attacked him. We tried to run them away, by hitting them with the sticks, but they wouldn't stop attacking the goat. It appears they were possessed, so I stopped hitting the dogs. I told my sister and brother to stop because I thought they might attack us.

They stopped hitting the dogs, but by that time, the goat was critically injured. His white fur had turned red from the goat's blood. He fell to the ground, and we ran back to the house to tell Mom and Dad what had happened. The goat belonged to Mr. Lee, and my dad informed him about what happened; he told dad to finish him off, so he did. Mr. Lee gave the goat to the family, so my dad and uncle strung him up in the tree and then started preparing the goat for consumption. I used to eat goat meat until this very day. After seeing how the dogs attacked him and to hear the goat's cry were enough for me. Besides, the meat smelled and tasted stinky; somewhat like the opossum, and that was the last time my sister, brother, and I went hunting with sticks.

By this time, I had a better knowledge of life and began to really observe my surroundings. I noticed how Dad treated Mom, and it wasn't good. He would go on a drunken binge on weekends; only on weekends, never during the week would he drink. He would argue, curse, and sometimes physically abuse her. I also noticed how my brother Fred Jr. would defend her which would lead to ever more violence my family had to suffer through. Mom's health wasn't good; she had a tumor on her brain that was caused by physical beatings to her head from my dad. It was too dangerous to remove, so we all lived with that. Every night, and sometimes during the day, Mom would have migraine headaches. She took medication for it, but that didn't help too much. She would scream during some nights because it was so severe. The family was dysfunctional with no consistent rules in place. My parents had non-parenting skills and only an elementary education. Dad was just emotionally absent; he seemed as if he didn't care about anyone or anything.

CHAPTER 2

Starting School and a Talk with My God

I started school in first grade when I was seven years old that's because of the month my birthday is on which is November. I attended Vincent Elementary School. I liked school; it gave me a better place to grow and learn new things. Everything was great until the second year; the children were mean and tried to bully me. I was shy and didn't have any friends at the time. I didn't meet any until about the end of my third year I met a couple of friends, and they were treated the same as I was treated.

Once my fourth grade teacher, Mrs. Griffith, visited our home. She wanted to talk to my parents about my quiet disposition in class and why I wasn't participating. Mom was there, but Dad wasn't. He was still at work. She couldn't tell my teacher what was wrong with me because I never told anyone, but the reason I was so shy in school is because I was embarrassed by things my dad did. It seemed like everyone knew what was going on in my family. Even if they didn't, I thought they did. I can remember my fifth grade teacher, Mr. Howard, being as mean to me as the children in class. My family may have had plenty of food to eat, but I didn't have all the pretty little ribbons and bows, new shoes, and dresses. I wore *hand-me-downs* from my sisters and some cute little dresses that were given to Mom sometimes from the lady she did housework for. But the teacher verbally talked down on me for that in front of the class about my possessions, as the so-called masters used to treat their slaves.

I didn't say or do anything at that time, but I did tell my parents about the situation later. I don't remember if they talked with the school about it, but I hope they did. We lived on the farm until I was ten. Before we moved, I had a strong knowledge of who my God was. I started praying out loud and asking Him to come into my life and protect me. I would ask why me? Why was I born into this family? He answered and said, "Why not you? When I created you, I made you special. I gave you the right armors to withstand all evils that will come your way and try to penetrate your sole. Everyone is not strong enough, but you are. I'm using you as an example of what others can achieve if you only believe and live a righteous life." That made me feel a bit better, and I tried to live that young life with God in my heart.

My brother Fred Jr. moved back with the family; he and his wife had separated. He worked in pulpwood and binged on alcohol every weekend just as my dad. It was only a weekend thing, but they would end it with a fight. They didn't drink alcohol during the week. On weekdays, they would go to work, they would come home and eat dinner, and then they would watch television and talk about things that happened during the day. They seemed rather polite to each other and the family. This was one of the reasons I couldn't understand why they couldn't act normal on the weekends. I assume it was the alcohol, but I'm not sure if they were alcoholics.

I believe alcoholics drink every day, but they drank only two days out of a week and couldn't handle it. I'm not a psychiatrist or psychologist, but I think they had mental problems, and alcohol probably made it worse. They never got diagnosed, but God I wish they would have. On weekends, they were like little children running around on the playground without supervision. My sister Betty left for New York on a job expedition. She had a son Robert Louis Jordan, who is four years younger than I am. He stayed with us while she was away; he was a handful.

We decided to move. Everyone was so excited about moving to another town. To us it seemed as if we were moving to a big city. We gathered up furniture, household goods, and clothing. We didn't have any more livestock or chicken, so the move wasn't too strenuous. We all said goodbye to our aunts, uncles, cousins, and a few friends. We were looking for a new phase in our lives. God knew I was. I hoped this change would make the family happier for us all. We wouldn't be living miles away from other families; the house we found was next door to other families. It was a large neighborhood and was like that for miles down the road in front and back of us on both sides of the road. I thought if we lived close to other people, Dad and Fred Jr. wouldn't fight anymore. They would be too ashamed to have others listening to the noise and nonsense they carried on. Not only did the commotion happened just on weekends but it would also happen on Friday and Saturday nights. During the week, there was peace.

CHAPTER 3

A Move from Farm to Town

We moved to another small town called Childersburg, Alabama. We seemed to be happy for a while. Mom; Dad; youngest brother, Denson; and I joined the neighborhood church, Enon Baptist. My brother and I were baptized; I was ten, and he was twelve. I enjoyed going to church with Mom and Dad being by my side. We were in church as a family for about a month. Then Dad stopped going with us. After a while, Mom stopped attending, and Denson left when Mom did, so that left just me. I continued to attend since it was within walking distance from our new home.

This new home was a bit better than the little cabin we left behind; there were five rooms and a big porch. That means we now have three bedrooms as opposed to one bedroom at the cabin. There was access to city water and not a well as we had at the cabin, but it came in the form of a hydrant located outside in the front yard. We still had to bring water inside using pails and buckets to cook, clean, and bathe. We still had an outhouse and a portable bathtub though. Most of the houses weren't built with plumbing pipes on Fayetteville Road, but as I said, it was a better place for a family. I used to love listening to Mom sing those old church hymns, while she was cleaning and cooking. She didn't always sing but would sometimes hum the tunes.

Dad started work for the city of Sylacauga which is another small town about seven miles east of Childersburg. It was a better-paying job. We could have used the money, but all of it didn't go to good use. Since he was binge drinking every weekend, a lot of the money went toward the liquor he drank. Life wasn't like living on the farm where we owned a few livestock and harvested fruits and vegetables to feed the family and we had to pay rent as opposed to getting free rent on the farm. Here we didn't have those resources and had to rely on Dad and Fred Jr. to make ends meet, but ends didn't always meet.

Dad stopped hunting because Uncle Tot had passed away, and other lands were privately owned, so there were times when we didn't have what we wanted or needed. I befriended some children in the neighborhood. First, there were the Wallace family who lived next door to us. The parents were Otis Victor (O. V.) and Barbara Wallace. The children were Nakita, Veronique, Bahri, and Adrianne Wallace. They treated me like I was a member of their family; they made me feel very special. Whenever they visited their relatives in Sylacauga, they would bring me along with them.

Two other neighborhood friends were Mary Jo and Joe Lewis Swain who lived across the road from my house. I used to love playing hide-and-seek, red light, tag, and racing with them. Their mom, Mrs. Mary Fluker, was married to my cousin, Mr. Prime Fluker. He was much older than she was and suddenly, passed away. She met a new boyfriend and would bring us to the movie theater sometimes along with him to see *The Pink Panther*, *The Jungle Book*, and a few other children movies. Most of the time, they would leave us there and come back to pick us up later.

Mrs. Mary would have a spring-cleaning day where they would wash all the windows in the house. I would help, and she would give me two dollars. That was a lot of money in the late sixties. I liked Mrs. Mary until one day Mary Jo and I walked to town about one-half mile to get something for her from the drugstore. As we were about to enter the store, I looked down and saw a ten-dollar bill lying on the ground between the curb and the pavement. I was afraid to pick it up. I was so shy until I thought the person who dropped it would see me pick it up, so I asked Mary Jo to pick it up, and she did.

We went to the Cohen's clothing store and divided the money. We both had five dollars each and bought a dress and a pair of sandals for three dollars total. The dresses were almost identical; they were red, yellow, and green plait, except mine was redder and Mary Jo's was greener, but the sandals were identical. They were white with thongs between the toes. We were so happy; we were going to dress alike when school started in the fall. As soon as we returned from town, we told Mrs. Mary about what we found, and the words that came from her mouth cut through my body like a sword.

She told us that we didn't find the money, we just sold our *you-know-what* to some men. Mind you, I was only twelve years old, and Mary Jo was just ten. This was at ten o'clock in the morning, and it only took about one hour to visit two stores and a round-trip walk to and from town. We didn't have boyfriends or men friends. Besides that, we were tomboys. I ran home with tears in my eyes and never went back there again. I told my family what happened; Mom and my sister told me not to worry, that's the way she operates with *her* body. She tried to lock us into her mind to be just like her. I still considered Mary Jo as a friend, and we still dressed alike sometimes in our dresses but didn't visit each other anymore.

Another good friend was Katie *Mae* Ware. She didn't live in my neighborhood; she lived in Fayetteville, like the road we lived on; that was the road you would take to get to her house. She, her sister Velma, and two youngest brothers, Robert and Cleve Ware, would walk to my house. Sometimes, my brother and I would walk them back home. Boy, that was a long walk. We didn't mind; we played all the way there. Sometimes we would spend the night. As I recall, once we stayed the night, got up the next morning, ate breakfast, and went swimming in a creek. Katie and her siblings and even my brother knew how to swim, but I didn't. I would walk across the creek with head held high, and I was good, except when one of her brothers dived in splashing water in my face causing me to fall backward. I couldn't get my composure and almost drowned.

The next thing I knew was Katie had pulled me out of the water and was giving me mouth-to-mouth resuscitation. I made her my BFF until she passed away in 2019 from a brain embolism. She had moved to Boston, Massachusetts, with her family when she was eleven, but we stayed in touch by phone and text.

Sometimes we would see each other when we came back to visit other relatives.

There were four other childhood friends whom I played with: two siblings Marilyn Carlisle, who was about thirteen, and Charles Carlisle, who was about eleven; and their cousin Carolyn Black was about fourteen.

I guess I can call them my friends after poking Charles in the face with my umbrella one day on the way home from school. We didn't catch the bus home. We lived very close to school, so sometimes we would walk home. Charles was picking on my nephew Robert Louis who was seven at the time. Robert may have been a handful, but I wasn't going to allow anyone to pick with him, so I told him to leave my nephew alone. He told me, "That's why your mama and daddy fights."

I really didn't know where that little speech fits in. I may have been a shy person prior to this incident, but I believe this brought me out. I hit him in the face with the umbrella, and his sister said, "Brenda, you didn't have to hit him in the face like that."

And I told her, "No one talks bad about my parents like that."

So we stopped speaking for a while but slowly started back. It wasn't the same though. I mention there were four other friends. The other one was John *Joe* Griffin who lived across the road from me. He was fourteen, and his parents lived out of town. So he lived with his aunt, Ms. Carson, who was a teacher. He would come to play with us sometimes. I had a crush on him, and when he found out, he wrote me a letter asking me for some tongue. I thought he was asking for something else, so the crush *flew out the window*.

CHAPTER 4

Attending a New School and Meeting Friends

When I started middle school at Phyllis Wheatley, the grades there were first through twelfth. I was in sixth grade, Mr. Russell Williams's class, and met new friends. It seemed that the children at Phyllis Wheatley were friendlier than those I encountered at W. A. Jones Elementary in Vincent. There were a few friends here whom I could visit, and they visited me. I remember one day, two of my friends, Deloris Rolston and Nita Trailor, came by my house and asked if I could come by their houses. I asked Mom if I could go, and she said to me; "Girl, you know you don't have to ask me if you can go, you know you can."

That wasn't what I wanted her to say; why couldn't she have said a simple yes? That remark was embarrassing to me because my friends were there and heard it. I wanted them to think my mom was a strict parent, and I was really showing her respect. From that day on, I just let her know where I was going and when I would be back. She was okay with that. I really loved my mom, and she loved me. I liked how she dressed; she only wore dresses. I was the same back then; I only word dresses also. I was a young adult before I started wearing pants or shorts. Now I switch from one to the other. I'm a feminine person, and dresses help me feel that way.

My brother Fred Jr. bought me a shiny new bike when I entered seventh grade; it was green and white. I enjoyed riding my bicycle and kept it locked in my room when I was in school or just wasn't there. Two of my siblings, Yvonnia and Denson, would pick the lock and ride my bike while I was in school and put it back before I returned home. I didn't know they had done that; no one told me until I became an adult. They told me then; it was more of a laughing matter. They said Mom would fall asleep, and they would sneak the bike out of the room. I spent my seventh-grade year in Mr. Stamp's class. That was one of my best years in school. I liked when we were at PE. We would use students as pyramids to run and flip over them. Most of the time, I would land on my feet.

Around the end of my seventh-grade year at Phyllis Wheatley, my class began to sell vegetable and flower seeds, and the ones who sold the most would become king or queen for the May Day event. I brought home two dozen packages of seeds. One pack was vegetables, and the other was flowers. Remember, Mom loved planting, so she bought them all. She told me to get one more pack of each. I did, and she bought those too. I didn't have a chance to sell any seeds to the neighbors. I won the sale for queen, and Mom bought me a beautiful pink-and-white dress with a tiny green bow at the neck area. Oh boy! I felt like a queen.

The winter of that year, my eldest sister, Annie Mae, and her husband, Henry, passed away in a house fire during winter of my seventh grade at Wheatley High School. That was a sad time for the family; it really took a toll on Mom, but we got through it by the hardest. We carried on ordinary lives for a while. Mom's health was slowly deteriorating. Dad stepped on a nail at work and didn't get it checked by a doctor. It later became infected and tried to manage it himself, but he was fighting a losing battle. I can't remember him seeing a doctor for any reason.

I attended Phyllis Wheatley for three years. I really can't remember my eighth grade teacher's name there. Maybe it's because of a tragedy that happened during that time. The schools had begun intergrading. My cousin Clarence Young, better known as *Kucoo*, another friend Eddie Jean Bahner, and I decided to transfer to Childersburg High School which was approximately two miles from my home and majority white. I had missed the deadline to register because I spent the summer in New Orleans with my sister Betty. This was the summer before my fourteenth birthday. I asked my parents if they could go to the school board and sign for me to register to transfer there. They agreed and went to sign the application for me to attend. I was so excited because I wanted to experience different cultures.

So the three of us attended Childersburg High School two years before the exodus of everyone else from Phyllis Wheatley. There were only about five blacks there. Two of them were Julius Twymon Sr. and his friend at the time whom he later married.

The transfer of students from Phyllis Wheatley was totally different from that after the closing and transfers. I will explain a bit more in details why this was so in the following chapter.

CHAPTER 5

Transfer to Another School

I was in the eighth grade when I stepped into Childersburg High School. I felt a different kind of welcoming; it was quieter. These were times when students got rowdy like when the teacher leaves the room for a few minutes, the boys would start a spitball fight; I really hated when the teacher left the room. I could see spitballs fly around the room like you could see bullets fly around in a Western movie. Luckily, none of them hit me, but some would land on my desk sometimes. I remember telling my family about these actions. My sister Yvonnia would ask, "What did you do?"

And I told her, "I raked them off with my arm and continue doing my classwork." I remember looking up at the ceiling seeing that it was filled with spitballs. Sometimes a few would fall at any time hitting students or landing on desks.

Besides the spitballs, I still enjoyed the school. Teachers there were very nice, and most of the students were also, except for those spitball throwers. I've never been bullied or called the *N* word at Childersburg High School. As I walked the hallways during restroom or recess, I would glare into other classes and noticed there were only one or two other people of color which was all right with me. I just wanted change, a good education, and a loving family structure at the time. I wasn't fortunate enough to get both, so I had to settle with one of the two. My father and his oldest son were still binging on liquor every weekend and with that, came fights when he starts to physically abuse my mother.

Fred Jr. would intervene, and that would be a hell of a fight. My brother was bigger and much stronger. He would overpower Jack in every fight. That would make Jack very angry, and he would run to get his shotgun every time. Fred would have to quickly run away and stay somewhere else overnight. The next day, they would act like nothing ever happened. It seemed ironic that they drank and fought only on weekends at night, and it only got worse. I was sort of glad that Fred protected our mother from abuse, but I couldn't stand those brutal fights.

After the first year at Childersburg High, I took a trip with my nephew Robert Louis Jordan to New Orleans; his mother Betty lived there. This was my first trip to New Orleans, but Robert's second trip. Remember, my mother, his grandmother, was his caregiver until he was about eight years old. My second oldest brother, Tommy Jene (Clay), took him to New Orleans once before. I was an excited fourteen-year-old on a Greyhound bus with my ten-year-old nephew. After arriving at the New Orleans's bus station, we took the RTA streetcar to Betty's house. She was a single mom of two smaller boys, Fernandez and Aaron Steele, who were about four and five. No one was home to let us in, but Robert knew how to get in another way from one of the windows. So he climbed up, opened the window, went in, and unlocked the door. Yay! We were in.

My sister came home late that day; she was at work. I got to see two more of my nephews whom I've never seen before, Fernandez and Aaron Steele. There were other nieces and nephews back home. My sister Georgia Ann Baker, we all called her Ann, was the mother of five boys and one girl at the time: Gary, Sammy, John, Curtis, Demetrius, and Sylvia Baker. She later gave birth to two more children, a boy, Melvin, and a girl, Sharron. My sister Marie Malone had three boys and two girls at the time: David, Tommy, Daryl, Sarah, and Angie. She later gave birth to two more children: Steven and Corey. My sister Yvonnia Chancellor didn't have any children at that time and still doesn't have any. Neither did any of my three brothers.

I had seen all of my nieces and nephews, but not the ones in New Orleans until now. So there I was, fourteen years old in *The Big Easy*. I loved that city so much, and Betty asked if I would like to stay and attend school. I called our mother and asked her if I could, and she said I could. We had lots of fun in *The Big Easy*; I met a few friends. One was Elaine Williams who was just a few months older than me. She and Robert helped me get familiar with parts of the city. Elaine and family lived in a house in front of my sister's three-room apartment. The living room and bedroom were combined, a kitchen and bath. Betty slept in the master bed, and we had rollaway beds that we would set up at night.

While Betty was at work, I would babysit the boys. This was the kind of life I could have gotten use to with all the convenience one needs in the city. I say could because I was only there approximately two months. I was supposed to start school in New Orleans that fall, but we got a call from the family in Alabama about a tragedy that just had happened one Saturday night. Fred Jr. had been killed. He was shot to death by the hands of our father during a drunken rampage. Jack was physically abusing our mother at home when no one was there to help her. Someone got in touch with Fred Jr., and he rushed home to save her. After he arrived, Jack shot at our mother, but Fred Jr. saved her, losing his life in the process. Our mother got away and ran to a neighbor's house, and Fred Jr. was the one who was shot in his neck, fatally killing him.

CHAPTER 6

Be Careful Who You Befriend and Missing Mom

On the return home, the first person we saw while pulling into the yard was our mother. She was standing outside in the front yard leaning against the side of the house. I will never forget the sound in Betty's voice when she saw her. She said, "Look at my poor Mom."

She was standing there as if in a trance. She didn't seem to notice us until we got out of the car and hugged her. She looked frail. This was the worst feeling I've ever had in my life. Domestic violence affects more than the couple involved. Often, the children are overlooked, unprotected, and left to endure the pain alone. That was me, I was a silent victim. I decided to stay in my hometown and continue school. I wanted to be near our mother because I really loved her. Jack still behaved the way he did before the tragedy, still abusing her, but now she didn't have that support she had with Fred Jr.

Yvonnia and Denson were never hardly home on weekend nights, that's when they went out. Most of the time, they stayed away all night, and sometimes they stayed away all weekend. So it was just me, still a child. Many nights, I tried to muzzle sounds by placing a pillow over my head, but that didn't help any. I just had to endure it. My mother's health had gotten worse and suffered a stroke. She had her children's support, and Jack seemed to start treating her a little better. At this time, she was very ill.

I returned to school in the fall of 1968. Things were strange. My friends tried to treat me the same as before the tragedy, but they just didn't visit. Who can blame them? But I was happy at school. No one mentioned anything or questioned me. I remember when I registered for the home economics class, I loved it and learned the dynamics of cooking and sewing. Teachers who helped to make a positive change in my life were Mrs. Armstrong, Mrs. Overton, Mrs. Limbaugh, and Mrs. Brewer. They taught me skills I should have learned before now but didn't. Advice from these teachers helped to start me on the right path. My favorite subjects were Language Arts, Biology, Home Economics, and Physical Education.

In Biology, I was intrigued by animal cells and how they divide and the process of mutation. In Language Arts, I liked how to put words together to form correct sentences, such as subject-verb agreements and how or when to use pronouns, adjectives, and adverbs. In Home Economics, I learned the basics of cooking, sewing, and a few other necessities. I was also fund of our PE classes. I learned patterns of how to exercise to help keep the body strong, including the eyes, face, gluteus, and pelvis, even the mind through sleep, meditation, puzzles, and a few other things.

I really liked gymnastics where students physically make pyramids with their bodies and another student would run from a distance and when close enough, jump and tumble in the air over the pyramid. Best scores were ones we received when we landed upright on our feet. I wasn't the best athlete, but I was good though. Best two were Bobbie Patterson and Bonnie Payne; I landed on my feet most of the time, but those two girls landed on their feet all the time. We also did track and field, soccer, volleyball, golf, high jump, and many other activities. Bobby and Bonnie were best in all sports among the groups. They were two hydrogenous Caucasian girls; they were smart and very nice to everyone.

I had a birthday and turned fifteen on November sixth that fall. I applied for student work program and landed my first job at Childersburg High School. It was typing and filing papers in the school's office. Those jobs were for students in low-income families. One of the teachers asked if I wanted to work part-time in the office two hours after per day after school. I said yes. The school's secretary helped complete the application, and they placed me in the office. I told the secretary I didn't know how to type well, and she said, "Do your best, I'm not looking for quantity but quality, so I did my best and liked it."

Before the job, I remember going downtown so I could buy supplies for a school project. I went to the five-and-dime store. I did not have enough money to buy all my school supplies. As I come out of the store, I noticed Mr. Funderburg, our landlord, standing by the side of the store. He was observing some of the activities going on downtown; he would do that a lot. I had bought all I needed except three poster boards, so I asked him if he could loan me a dollar to buy them.

He said, "No, your daddy is already behind on the rent."

Poster boards back then only cost about twenty-five cents each. How could someone deny a child who needed something so simple from enjoying that privilege in school?

Fred Jr. used to buy my school supplies before his death. My father did give me a few dollars, but it wasn't enough. I ended up getting it from my sister Ann and then returned to the store and purchase the items. When I started working, I brought home seventy dollars per week and bought all my school supplies and clothes. I tried to give mother a few dollars, but she wouldn't take it. She said I needed it more.

I worked in the school's office approximately two years. Our mother was still having those awful migraines and seemed to have had a mild case of dementia. Yvonnia and I took care of her as much as we could. Our father would still binge some weekends but not as often. He was still having a problem with his foot. Previously, I mentioned he had stepped on a nail at work and didn't get any help. It became infected, so he decided to go see a doctor. He waited too long; he found out it had caused gangrene in his foot. He was still working at the time but walking with a limp. Our mother ended up having a stroke. It didn't leave her paralyzed, but her mind was getting worse. She didn't talk as much as she used to.

I was walking to town one Saturday morning and met someone named Larry Hamilton. This was at the beginning of my eleventh-grade year; I was sixteen. We became friends, and I introduced him to the family. He knew most of them already. I assume by them being out at night going to juke joints. I wasn't going out at the time and didn't drink. We dated a couple of months before my mother had another stroke. She was in the hospital a few days and wanted to see all her children. We were there one day, and she talked with us. Everyone went to the cafeteria later to get food, but I stayed there with her. She knew, and we knew she was about to grow her wings. She talked with me about how badly Jack had treated her. Then she said, "Brenda, I don't know what else to tell you, except to do whatever you think is right and keep Yvonnia away from Jack."

She and Jack didn't get along either because she tried to help Fred Jr. protect our mother. Those were the last words she would ever say to me.

Some of us returned home that day, but Betty and Yvonnia stayed with her. I prayed she would get better. I didn't want to lose her; didn't know if I could live without her. Later that afternoon, we got a call from Betty at the hospital. Mom had finished growing her wings. I felt that was the worse day of my life because I knew I would truly miss her. We weren't as supportive with each other as some families would have been. No one talked with me about *the birds and the bees*. I had to learn on my own, the hard way. Shortly after Mother died, I became pregnant. To me that was another devastation. I felt alone, and then I thought, *At least I have my friend Larry*. Be careful who you befriend. He didn't show any compassion or care. Later I found out he was a Casanova. He had dated almost every woman in that small town.

I continued my classes throughout the fall, but at the beginning of the new year, I became pregnant and dropped out of school. My cousin Clarence Young, we called him Kucoo, always watched out for me and told me that Mrs. Brewer, my eleventh grade teacher, said if I come in and take the end of the year exam and pass, I would get promoted to the twelfth grade. School would be out in a couple of months, so I took the test and passed. Meanwhile, Larry and I were married. When the baby was born, we named him Larry Jr. after his father. He moved to Atlanta where he found work with his Uncle Jeff. I joined him later while Pearl Hamilton, Larry's mother, kept the baby.

It was a sad time there; we didn't go out or do anything together, but Larry was out all the time; I barely saw him day or night. His cousins were very nice to me. I visited them almost every day. One day, I visited them, and there was another young lady there whom I hadn't met. Two of his female cousins told me that was the girl he was trying to date. I confronted him, and of course, he denied it, but

I knew better; that was in his nature. During this time, he never took me out anywhere, not even to a fast-food restaurant. Except for this one disgusting place which I wouldn't even call going out, it was an X-rated movie which I didn't even enjoy. I figured our first time out would be something special like a regular movie, concert, or dinner.

We were in Atlanta for a couple of months; Larry's mother figured we had settled in and brought the baby to us. We were in Atlanta almost a year, and then moved to Fort Myers, Florida, where he had other family members: Brother Leonard, Sister Ozel, Aunt Daisy, several cousins, and a sister-in-law whom I can't remember the name.

There was another sister who lived at home with their mother Pearl. It was twelve hours before we arrived in Fort Myers from Atlanta. The car started giving us trouble. As soon as Larry turned the ignition off, it never started again, and he never got another one. We stayed with his brother Leonard and his wife about two weeks, and then we started renting our own home. We didn't have a car or a phone. I had to walk to the store when I needed food or just to use the phone. It was here when he really started to treat me badly. He started slapping me around and neglecting his son and me. I didn't know anyone there, except his family. I felt like I was in prison, and he was the warden.

His sister-in-law who worked at an orange orchard let him know that there was a tomato farm and they needed people to work, so he started working there. Then one day, he came home and told me that I could work there also. He found a nursery in the neighborhood, and a bus would pick us up not far from there. He worked about two weeks and quit, but I continued working. That's when he started staying away from home days, up to a week sometimes.

In a short while, Larry Jr., who was only five months old, was very sick. I thought it was a cold and tried to treat it. In a few days, it got worst. He had a fever, and my husband was nowhere to be found. So I called a taxi and brought him to the hospital. He was there a few days before doctors found out what the problem was. He slipped into a coma. They found out it was meningitis, a highly contagious disease. Doctors didn't know too much about this disease in nineteen seventy-one. It was a rare disease and according to his doctors. There wasn't enough information about it at that time.

Larry came to see him once after a week in the hospital. My son stayed in a coma for two weeks. They said they didn't think he would make it, but he did. They also said that he would probably have some brain damage, and that was true also. He was late doing everything and unable to communicate verbally. He also had seizures, but the medication they gave him kept it under control. After my son was out of the hospital, I let the nursery know what was going on. They told me he wasn't the only baby who contracted the disease, there were two or three other babies who was sick prior to my son, and he probably contracted it from being there.

I wasn't happy there in Fort Myers, so I called my sister Betty in New Orleans and asked her to send me a one-way ticket to come live with her for a while. She said yes and sent it the next week. Larry Jr. was out of the hospital, so I packed a few clothes and called a taxi to bring me to the Greyhound Bus Station in Fort Myers. While waiting for the taxi, my husband Larry came home and looked at the ticket. I thought he was going to rip it, but he didn't. He left instead. The taxi came, and I prayed he wouldn't follow and make me get off the bus. He didn't, and hallelujah, I was on my way to New Orleans with my son. I lived in Fort Myers about six months and was with Larry for six months before I left. I was so happy to be rid of another demon in my life; the first was moving away from my father. Now I was past due for a new beginning away from my husband, Larry Hamilton.

CHAPTER 7

A Temporary Move to New Orleans, Louisiana

My son and I arrived at the bus station in New Orleans. It took twelve hours to get from Fort Myers to New Orleans on the bus. Robert and Betty's friend picked me up from the bus station and brought me to Betty's house. She had moved from the one-bedroom apartment to a bigger two-bedroom apartment on South Dorgenois Street. I stayed there with Betty and her three sons: Robert, Fernandez, and Aaron. I started working with her at the hotel and later, met my friend Elaine again. Elaine started babysitting for me. Betty met a new friend Preston Cox on the RTA bus going and coming from work. I was eighteen years old and started going out for the first time. I was eighteen years old and as *green as grass* but held my own. From what I had come through, I was an independent young lady who carried myself in that fashion. I vowed to never let anyone ever again use me as my husband had done. That was a lesson in disguise.

We received a call from Yvonnia in Childersburg one Friday night. She informed us that our family home had burned down. My father who had recently retired was alone and on a drunken spree and thought his son Fred Jr. was in the house. Neighbors said that just before the fire, they heard him yelling Fred Jr.'s name as if he was there and torched the house. No one was hurt, and that was a blessing.

I lived in New Orleans for eleven months and enjoyed being there. Never will I forget the first musical fest I attended; it was *Soul Bowl '72*. I was eighteen, and Robert was fourteen. We were there at the Tad Gormley Stadium in city park. I remember his arm being in a swing. It was broken from playing basketball in high school. We rode the bus to the park, and it was jam-packed. People were brushing against Robert's arm, while I was trying to protect it. We had lots of fun; we saw artists, such as Jackie Wilson, Ann Peebles, and a few others. I attended lots of events with family in New Orleans. I was there a few months and then wanted to take a trip to visit family in Alabama, so my son and I boarded the Greyhound bus once again.

It took twelve hours to travel from New Orleans, Louisiana, to Birmingham, Alabama. Yvonnia and her friend came to pick us up because the Greyhound bus didn't travel to Childersburg, which is about thirty-five miles from Birmingham. The only bus service that traveled to Childersburg was Trailways. Later I was in Childersburg reminiscing with family and friends. I was staying with my sister Georgia Ann and her family. Our father was staying there also because the family's home didn't exist anymore after the fire.

My plan was to remain there for two weeks, but life didn't always go as planned. I ended up calling sister Betty and informed her I was planning to move back home. The reason was the living conditions at my sister's home in Childersburg. There were eight children, a husband and wife, our father, and myself which is a total of twelve people in a two-bedroom house. I was sitting on the porch one afternoon and saw a car pass by. It stopped and backed up and turned into our yard. I didn't know who it was until he stepped out the car. It was Wylinsky Keith, a friend I rode the school bus with in the past. Everyone called him Willie. He and his brother Henry would always save a seat for me on the bus, and that made me feel very special.

After the encounter on the porch that day, he asked me out on a date, I said yes. Willie was working in construction building the latest development of project houses in Childersburg. He said some units were to open soon, and if I were to stay there, I should apply for one. I talked with our father and told him the situation, and we applied for one. His foot had gotten worse, and his doctors told him they may have to amputate some of his toes. I didn't know if I wanted to share a home again, but life is about learning, decisions, and forgiveness. People can change, but at the same time, we're all creatures of habit. They say resentment can cause a heavy heart. I forgave my father and was able to release the pain I carried since early childhood, and we got the apartment we applied for.

CHAPTER 8

My Knight in Shining Armor

Meanwhile, I was with sister Ann for approximately one month. This was winter season, and Willie and I had grown serious in the relationship. We had dated for a while and then found out I was pregnant with a second child. My sister had a wood-burning heater in her house. Some mornings, I would hear someone chopping wood at the woodpile. I walked to the widow and saw it was Willie. He chopped the wood, then came inside to build a fire in the heater, and went the work. That was when my brother-in-law was still at work. He worked a rotating shift and had not made it home yet.

Later Willie landed a job with Alabama Power in Wilsonville, Alabama. Meanwhile, my father and I moved into our brand-new project home. We were the first family to live in the apartment. Yvonnia and Denson moved in with us. During this time, my second child, Sharette, was born on August 1, 1973. Two years later, Willie and I were married in 1975. He found a house for us to buy in Killough Heights, Childersburg. He brought me to see the house, and it was lovely. I owned my own home at the age of twenty-one; Willie was twenty-three. He also bought me a nice, used red-and-black Pontiac.

We were happy in our new home. After a year, I found a job working at White Knight Manufacturing company. We manufactured hospital supplies, such as masks, hairnets, gowns, and gauze. On weekends, we would do shopping, dinner, and movies all in one day. That was a total difference from the first marriage. On day while shopping, we bought a brand-new Grand Marquis and added a double-car garage to our home. During the second year on the job, I was expecting my third child, Deitrich. I continued working until the seventh month of my pregnancy.

The doctor wanted me to stay of my feet because of overexerting myself. I went to the hospital and found I had dilated four centimeters; the maximum is ten. So I followed the doctor's advice, and within two months, Deitrich was born. After two months leave, I returned to work. We updated our home with better furniture. It seemed that every time we bought something new to better our lives, it seemed like we would get punished for it; people would gossip and say we didn't need that. I was just getting my husband in debt. Excuse me, but wasn't I working too?

In such a small town, people would try to stir up trouble when they see someone living well. We both had good jobs bringing in over sixteen hundred dollars biweekly, and that was the seventies. I didn't see what the problem was at first but later realized they didn't have anything else to do in that small town, so when they were bored, they gossiped. They stayed in others' business and couldn't navigate their own. That's one reason I began to not like living there anymore.

Shortly after Deitrich was born, Larry Jr.'s grandmother, Ms. Pearl Hamilton, my ex-husband's mother, was retired and agreed to care for him after we talked about my busy life with three children, work, and a husband. Larry Jr. had special needs. She knew him better than anyone and was in for the cause. She lived only ten minutes from us, so we saw them often. Sometimes I would stop by to see how my father was doing. He told me his doctors said they would have to amputate some of his toes on his bad foot. He had stopped drinking alcohol, but it seemed it was too late; the damage was done. Gangrene spread to his foot, and it was amputated. Then finally, his whole leg had to get amputated. He moved in with his sister Aunt Loree Chancellor. She and her twin sons, Joyth and Royth Chancellor, cared for him. After a long suffering, he passed.

Yvonnia and Denson stayed in Childersburg for a while and then moved to New Orleans. I had mentioned earlier that two of my husband's hobbies were fishing and hunting. I remember coming home from work on September 18, 1979. My husband Willie had already come home from work and took the children to a sitter. I noticed he left a note letting me know he'd gone fishing and left the kitchen door open, but the screen was locked. About an hour at home, I received a call from Georgia Ann, my sister, asking if I knew where my husband was. I told her he went fishing because he left a note saying, "I went fishing." I still have the note. It survived Hurricane Katrina. It was in a bag with other important papers. Ann told me he was lost, and people were trying to find him.

I knew what that meant and started crying. Ann came to pick me up. It was getting dark. We picked the children up and then went across the river to Wilsonville. Willie was fishing behind Alabama Power Steam Plant where he worked. He was with a couple of friends. According to what I was told, he fell in the river while catching some minnows with his fishnet to fish with. The net had a long handle; it broke, and he fell into a current in the river coming from the steam plant. Some people asked if he could swim; I've never seen or asked him if he could swim. We never talked about it. But I do know that when he came home and left the note, he didn't change his shoes. He was still wearing his work clothes. The boots had steel in the toes, and that could be the reason he drowned, or it could have been the current itself or both. "Peace be still" (Mark 4:39 KJV).

He was found approximately two hours after he fell into the river. It was late evening, early night when he was found. I got out of the car to see him, but his brother, Henry Keith, told me I didn't want to see him in the condition he was in. I did want to remember him the way he was, and I thanked Henry for advising me. I stay in Childersburg about two years after Willie's death. We really missed him. God sent him to me for a reason, and that was to help pull me out of the storm I was going through.

Our daughter Sharette was six when he passed. She was devastated, and so was I; our son Deitrich was only two and didn't remember much. He was connected through pictures and conversations. My husband was twenty-seven years old when he passed, and I was twenty-five. My son Larry Jr. was eight, but of course, he was with his grandmother. Willie and I had started renovating our home; we added a double garage before he passed. I continued to by extending it to a two-level home. There were a large family room, patio, and a barbeque grill outside at the back that was built in the shape of a fireplace with bricks that I added downstairs. Upstairs, there were a large bedroom with a balcony, a bathroom, and lots of closet space. I asked God to guide me in everything I did in caring for my family as a single mother.

CHAPTER 9

A Permanent Move to New Orleans Louisiana

Shortly after I finished renovating my home, I accepted a two-week voluntary layoff and spent most of my time visiting my sister in New Orleans. I returned to Childersburg and met my friend Katie Ware's cousin; his name was Charles Ware. He was a nice guy who knew how to treat a lady. We dated for a while, but I realized we weren't compatible. Besides, he was already in a relationship and had a son who was the age with my son Deitrich; they were two years old. He was a sweet little fellow, but I didn't want to separate a child's mother and father, so that relationship didn't go too far. I adored his family, Mr. Milton and Mrs. Josie Ware. The siblings were very nice to me also. Before the relationship with Charles, I knew them because Mr. Milton Ware, who's passed away, was my sister Georgia Ann's uncle by law. Katie Ware, who's also passed, was Mr. Milton's niece and my Georgia Ann's sister-in-law.

I never returned to work because I had been contemplating moving to New Orleans. It started to seem like Childersburg was getting too small for me. I said that I was planning to move; I didn't get much static from my son Deitrich, but my daughter Sharette was highly angry with me. Eventually, I let other family members and friends know about my plans, and they too were just as angry. But I didn't let that stop me because I knew I would be visiting quite often. My husband was no longer with us, and I knew I've always wanted change. I couldn't get it there, so that was something I had to do for myself and for my children.

I didn't move right away, I needed to get a few things in order first. The most important thing was to find a decent place to live with the children in New Orleans. Betty volunteered to look for a three-bedroom apartment in the eastern part of the city, the suburbs, and sent the information to me. It wasn't long before she contacted me with the information I needed, so the children and I went to New Orleans to visit the apartment complex, choose one, and pay necessary fees. After that, we celebrated with members of the family in New Orleans for us moving there. They were happy for me, but back home, there were anger and sadness. I felt a bit of sadness that I was leaving family members and friends, but when I thought about the quality of life I would have in the city, the happiness would override the sadness.

I returned home approximately one week; I still had business to take care of there, so I contacted a real-estate agency to handle my home as rental property. After all my business was completed, I had some time to spend with friends and relatives. I let them know that there will be times for a nice visit from both sides. Time narrowed down for us to leave, so I asked my nephew Robert, a friend Lawrence

Spain, and another nephew Sammy Baker to help me move and drive the moving truck to New Orleans.

They came to help, and a few neighbors also helped. I remember as we finished packing, we were ready for the highway, but it took some time to get into the vehicles. The neighbors were standing around to see us off, but the children didn't want to get in the car. It took a while, but when they did cooperate, we were off to Louisiana. I drove the corvette; Robert droved the moving truck with Sammy as a passenger, and Lawrence drove the Grand Marquis. Upon arrival at our destination, we unloaded the vehicles and returned the moving truck to its destination in New Orleans.

Sammy stayed with us a couple of days and enjoyed the city, and then we brought him back home. We returned to my new home, and then it was time to get ourselves together. I contacted the schools the children would be attending and registered them. Robert and Lawrence drove me around the city to help me become familiar with the streets. Since I moved eleven miles from downtown, I was focused on ground-level streets in case there were a traffic jam on the interstate. In a short while, I started driving around all by myself, and that was the best experience.

It was 1981 when I moved from Childersburg, Alabama, to New Orleans Louisiana, and in 1983, the real-estate company contacted me to let me know the tenants had moved from my rental property without letting them know. They kept the last month's rent and took the remote control to the garage with them. I wasn't going to deal with that, so I listed the property with the real-estate company to sell the house. It sold immediately. Besides, I was paying rent in New Orleans, and each year, the rent would go up. Each time I would move to a cheaper apartment. Once I moved three times in one year. After the house sold, I bought a new home that wasn't too far from the apartment building where we lived. At that time, the children weren't angry with me anymore for moving from Childersburg; they had gotten used to New Orleans. In fact, they were happy after attending school and meeting the neighborhood children.

CHAPTER 10

My Career Decision

Lawrence and I dated a few years. After Sharette graduated from high school and went to college, I returned to school and received a general equivalent diploma. A lot of students were dropped out of school just to get this diploma and graduated the fast way just to get a quick start on life, but I wouldn't recommend that to anyone, especially if they don't attend college. They're missing out on a lot of pertinent information. So I enrolled at a community college and received an associate degree in business. I did receptionist/secretarial work for a couple of years. I wasn't satisfied with office work, so I applied to the University of New Orleans as an undecided student. Within a year, I decided that I wanted to do something to contribute to society, so I majored in early childhood education.

Lawrence and I were still dating and were talking about marriage, but I was reluctant because I had been married twice to his never-being married. He even mentioned that my house would never become his. Of course not, I had been a homeowner for eight years before I met him and managed to contain my livelihood without any problems. Lawrence's mother passed while he was in high school and dropped out. When we met, he was working for Barto Marble Company. We started dating around that time in 1981 and dated for eleven years until 1992. We didn't spend much time together because we were two busy people. I was in college full-time, worked a full-time and a part-time job, and had children who depended on me. I thought about it and gave him an ultimatum. If he would get his GED, then I would marry him.

He did return to school but only went for two weeks and dropped out. I figured if I was in college, holding two jobs and raising children, then he shouldn't mind getting a GED. He owned a floor cleaning service, but I knew from experience that when you drop out of school, you miss out on a lot of valuable information needed in life. One could make that up during adult school and college. That is needed for anyone owning a business to be able to communicate fluently with customers and other businesses. Although I attended a community college after getting a GED, receiving an associate degree in business, and working as a receptionist/secretary, I made a career change majoring in elementary education at the University of New Orleans. I was required to ELA and Math remedial courses for one year before the four-year journey could start. That means, I spent five years in an elite college, and it paid off. I received a bachelor's degree in early childhood education and a Louisiana state teacher's license.

After Lawrence and I had a serious talk, we parted our ways. I was a freshman at the time, and things were tough. I kept my head up and asked God to take me into

His hands and keep me grounded throughout the process, and He did. While in college, I made the state dean's list, the national dean's list, and graduated with a 3.7 GPA. After graduation, the first school I worked at as a certified teacher was Chester Elementary on South Dorgenois and Erato Street. I taught second grade for two years and kindergarten for approximately eight years until 2005. That's when Hurricane Katrina bared down her evil neck, and the whole city was dispersed in water, and residents were scattered around the entire United States.

My whole family in New Orleans and I evacuated to Childersburg. We returned home within six weeks and lived in a trailer provided by FEMA with my daughter, son, and first grandson, Javoughn Perkins. I also started back teaching, not in Orleans Parish but in Metairie, Louisiana, which is Jefferson Parish. A lot of friends and colleagues didn't return to their homes; a lot of them had started a life other than New Orleans. I was ten years into my career around this time. I wanted to come back because I had made New Orleans home and wasn't ready just to abandon it. I wanted to stay and fulfill the promise I made to the little children in Louisiana, and that was to teach, mold them, and help make a difference in their young lives.

CHAPTER 11

Looking Forward to Retirement

I taught fifteen years in Metairie, Louisiana, following Hurricane Katrina. Orleans Parish schools had not opened yet because of the severe damage all the buildings were in. I was away for six months and wanted to renovate our home. As I mentioned earlier, I stayed in a FEMA trailer with the family. To apply for a trailer, one would have to own some property or be a city or state worker. My daughter had to return two weeks prior to my return. New Orleans had more damage than any city or town in Louisiana. She couldn't live there, but Betty's daughter Takesha lived in Metairie and didn't get much water. Metairie is 6.2 miles away from New Orleans, so she moved in with Takesha temporarily because state employers mandated their employees to return to work after four weeks. They both worked in Jefferson Parish.

Most of the family stayed with Yvonnia after evacuating to Childersburg, Alabama. Not only were there family but Betty brought along two friends from her work. My sister's house was jam-packed, but not for long. As days passed, some started to leave, and after six weeks, we all had left to go back home or go live with relatives. I had planned to retire after twenty-two years of teaching, but after a conference with a representative of the teacher retirement system of Louisiana, I needed to teach three more years to get full retirement benefits, so I worked twenty-five years and was slated to retire in 2020. My friend and colleague Joyce Byrd was retiring at that time, and in 2019, we started to joke about it; we said in 2020 we're going out with a *bang*. Well, this is where the phrase *be careful what your ask for* comes in. We went out with a *boom*, not a *bang*. It was said that around the end of 2019 was the time when COVID-

19 entered the United States, and schools started shutting down in March 2020. That was our *bang*.

We started remote teaching the rest of the school year, and it came with problems: the class I taught was many ESL students, and a lot of them didn't have computers. Not only did they not have access to computers, but a few regular students didn't have access to computers either. The school loaned them computers, but unfortunately, we couldn't get in touch with all the parents to get computers to them. Out of twenty-four students, I only taught fifteen remotely. The school held classes from pre-K through eighth grade, and each class had about that many, give or take. A lot of parents didn't speak English, so that caused a lot of children to fall behind in their work. I'm not speaking only about Louisiana but schools across the United States.

Approximately ten teachers from my school, J. C. Ellis in Metairie, Louisiana, retired that in 2020. Most of weren't planning to, but circumstances took a toll. We did the best we could with limited resources. And finally, the last day of school arrived for me and a few other teachers that year. The first year of my retirement was not fun at all. Instead of me cruising to other countries, there I was walking around with face masks on like everyone else, afraid to touch anyone; well, you know the drill.

About a year later, scientists worked together to come up with a vaccine to help control the virus. Three years later, the virus is still around, but if everyone would follow the protocol and practice safety procedures, that would eliminate some cases and help control the pandemic. I'm up-to-date with all vaccines, not just COVID-19 but all others. I've never contracted the virus; at least I don't think I have because I never had the symptoms, but of course, one can be asymptomatic. Regardless, God has carried my family and me through this far, and I pray He continues.

CHAPTER 12

Living the Life My God Designed for Me

Considering the ups and downs of society, I don't see things as happening to me but happening through me. Meaning that I try not to get stuck on things I can't control. I just develop an opinion and ask God to take care of it and give me the courage to challenge things I can change. Dwelling on things you can't change can cause one to become offensive and angry, which can cause stress and that can lead to depression and mental illness. I start my daily prayer asking God to help me preserve my mind because that is the basis of keeping everything else in order. I don't just pray and ask God for help, but I partake in exercises and events that are appropriate for strengthening the mind, such as doing puzzles and exercising, especially yoga. My family and I visit city park to rent and ride bikes.

Another event I find relaxing is decorating; that is my hobby. Since retirement, I'm staying active doing activities that are essential for living a blessed life. I start the mornings in prayer and meditation. Three to four times, I'll do a total body workout. Later, I'll eat breakfast or brunch. Basically, I'm a late-eater in the early morning hours, so brunch works for me. During the day, there are household chores and errands to tackle. I'm not saying this because it keeps me on my feet. Movement is essential in the good health repertoire.

I'm working to keep my mind and body strong. I help plan exciting events with my children and grandchildren. Sometimes we'll rent a condominium on a Florida's beach. We rent cabins on campgrounds, fish, and ride peddle boats, flatboats and canoes. We barbeque and make s'mores. We visit relatives in Alabama, rent and ride bikes in city park, and host parties with relatives and friends. I like doing these events to stay active during my retirement and stay close with family.

This novel is for those who are going through troubles in life, no matter what they are, if you're targeting yourself or if someone else is targeting you, whether it's physically, mentally, spiritually, or financially. The focus is on those who are from lower-socioeconomic backgrounds, dysfunctional families, and families suffering from addictions or mental health problems. Many people think they're failures, but the only thing that will beat a failure is a try. If you're trying to change your life, and it seems difficult, just change your mind and try God. It's as simple as that. People fail because they are afraid to go through the process. They may think what they want will never happen, but remember, God hasn't forgotten you. We're born with His spirit inside of us and promise to forgive and take care of us if we repent; "ask and it shall be given you" (Matthew 7:7–8).

We often forget, lose it along the way, or have never read the Bible before, but He will forgive us. If we want good things to happen in our lives, we'll have to activate that spirit and change our way of living for the better. Never stop believing in our Lord Jesus Christ. There's an old saying that you're born with nothing, and you leave this earth with nothing, but I challenge that. I was born into this world with Christ in my heart and plan on leaving this world with Him still there. I can witness to the words in one of Aretha Louise Franklin's songs, "I was lost, but now I'm found. Was blind, but now I see" (1972).

Brenda and husband Wylinsky.

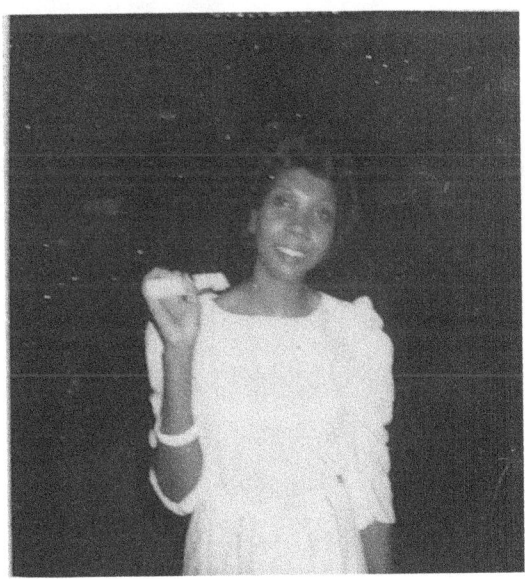

Brenda received a GED certificate.

Brenda, Yvonnia, Betty, and Marie

EPILOGUE

There are different seasons of life where one must deal with or manage good in life as well as bad. Our roads wouldn't be so rocky if we would invite God to walk along with us and allow Him to lead the way. He won't divorce you if we lose the right path, but He will pick you up and carry you. I blamed both my father and brother for our family's dysfunctions, but I've learned to forgive them because they were sick and needed help. Back in those days, society didn't recognize this as mental illness but as alcoholics.

I prayed for them to get help, but I imagined they need to prey for themselves because the only one that can initiate you being saved would be you yourself. Prayer helps especially if a lot of people is praying for the same cause. So "Children of God, love and forgive" (Daniel 9:9 NIV); you will be blessed. I found that everything happens for a reason. As a child, I used to ask God why was I born to this dysfunctional family with so much of His love and knowledge in me. Later in life, I realize that one doesn't question God's work; that was His plan. He takes the poor meek souls and mold them into his best work.

At the age of twelve, I made a promise to God and myself that when I have children, I won't allow them to be raised in the kind of environment that I was raised in. I wanted more than a three-room cabin with no inside water and an outhouse. So here I stand today, fulfilling that promise. My children were fortunate enough to be raised in a loving, happy, and safe environment where they can thrive and voice their opinions about anything. Life is a series of choices, and this one is mine. "With man this is impossible, but with God, all things are possible" (Matthew 19:26 NIV).

FINDING MYSELF IN THE STORM

Sharette and Deitrich

Larry D. Hamilton Jr.

BRENDA F. KEITH

Brenda, Kindergarten Teacher.

ABOUT THE AUTHOR

As a child, Brenda was curious about people in society and their behaviors. She always associated herself with those who showed empathy toward others. She's a dedicated and compassionate young lady, with a passion to teach small children and help mold them into their best selves into adulthood.

As a child of God, Brenda believes that prayer weighs heavily on miracles and spends each day in prayer, thanking God for the uncharted miles she has conquered. Brenda grew up on a farm with eight other siblings where she had a love for nature and countryside. Brenda was quiet and shy as a child in the country but loved helping her mother do chores, especially planting fruits, vegetables, and beautifying the yard with flowers. She often watched how people interacted with each other and quickly developed a sense of good versus evil.

Around twelve years old, Brenda developed a strong sense of God; thanks to her mother for bringing her to Bible study quiet often and introducing her to God. Brenda considered the morals and values she learned through experience and Bible study. Coming from a dysfunctional family, she is now an intellect who doesn't mind giving advice to anyone who will listen. After retiring, Brenda enjoys the pleasure of family and friends and the passion of writing, to continue to inspire and motivate those with a similar background.

In Loving Memory of Yvonnia Chancellor

Dec. 11, 1949

Sept. 24, 2023

Printed in the USA
CPSIA information can be obtained
at www.ICGtesting.com
LVHW050300290124
769813LV00051B/989